ABSTRACT BUTTERFLIES GRAYSCALE ADULT COLORING BOOK

A Galleria Monte
Art Project

COPYRIGHT 2016
GALLERIA MONTE ART WORKS
ALL RIGHT RESERVED

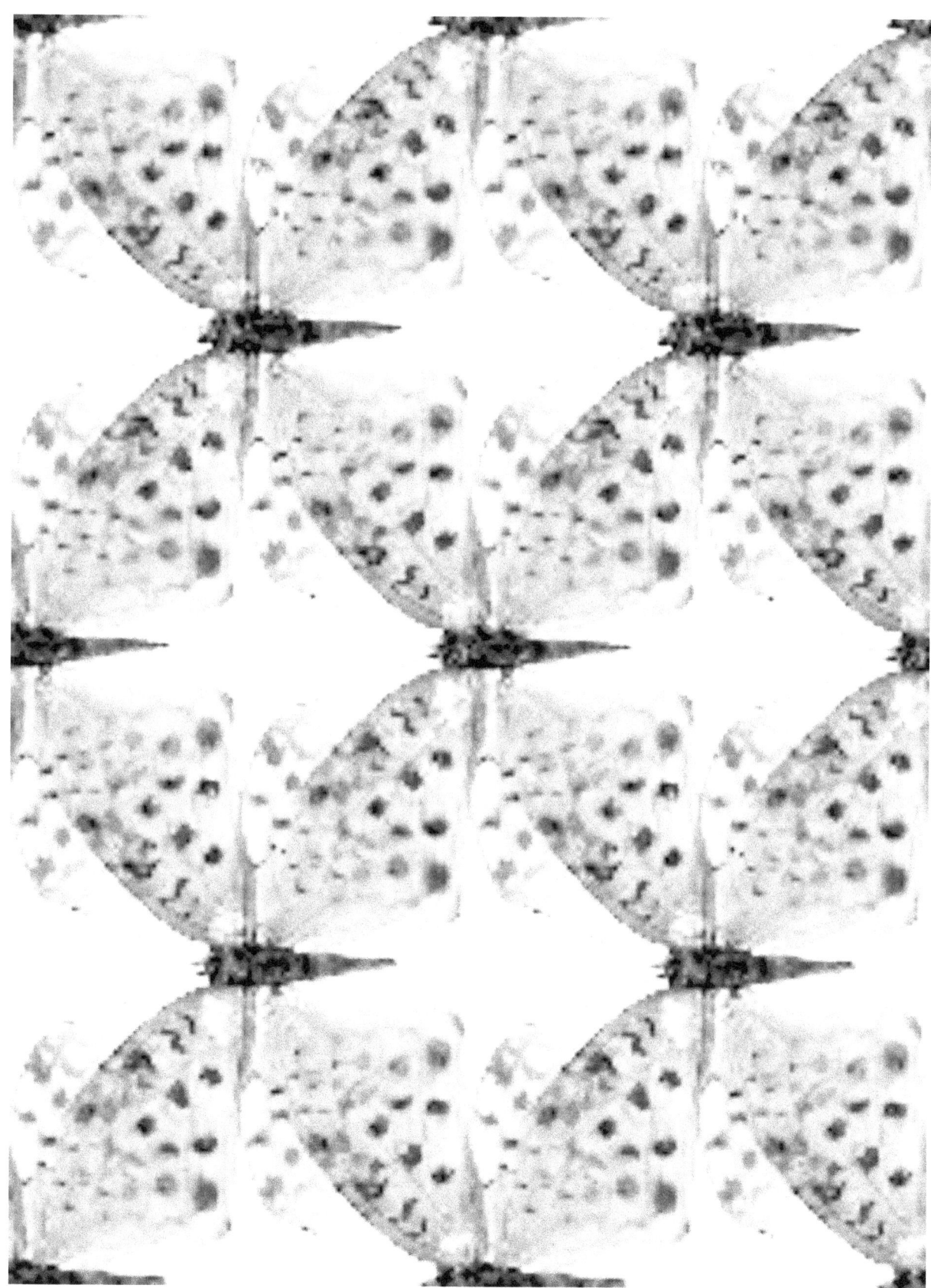

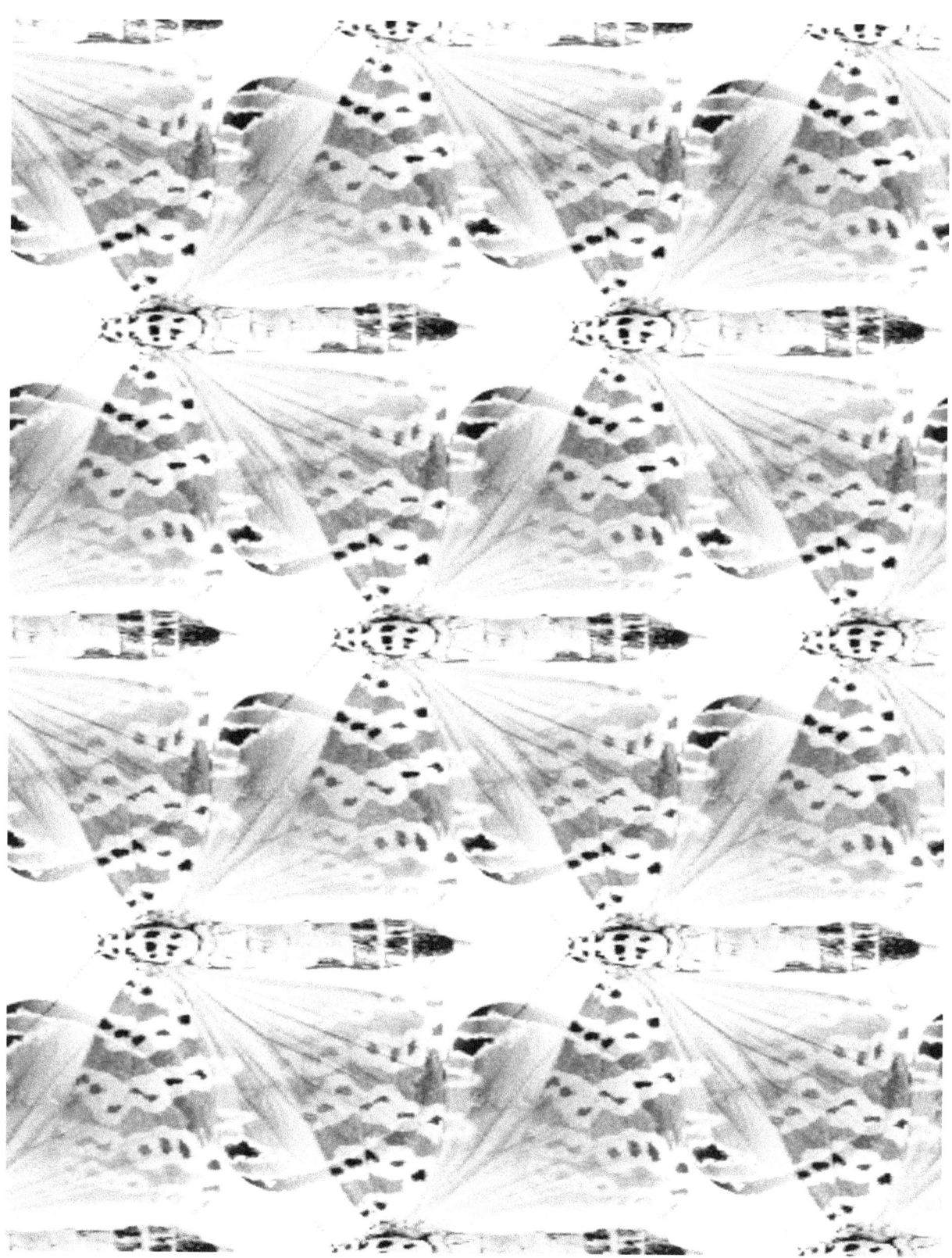

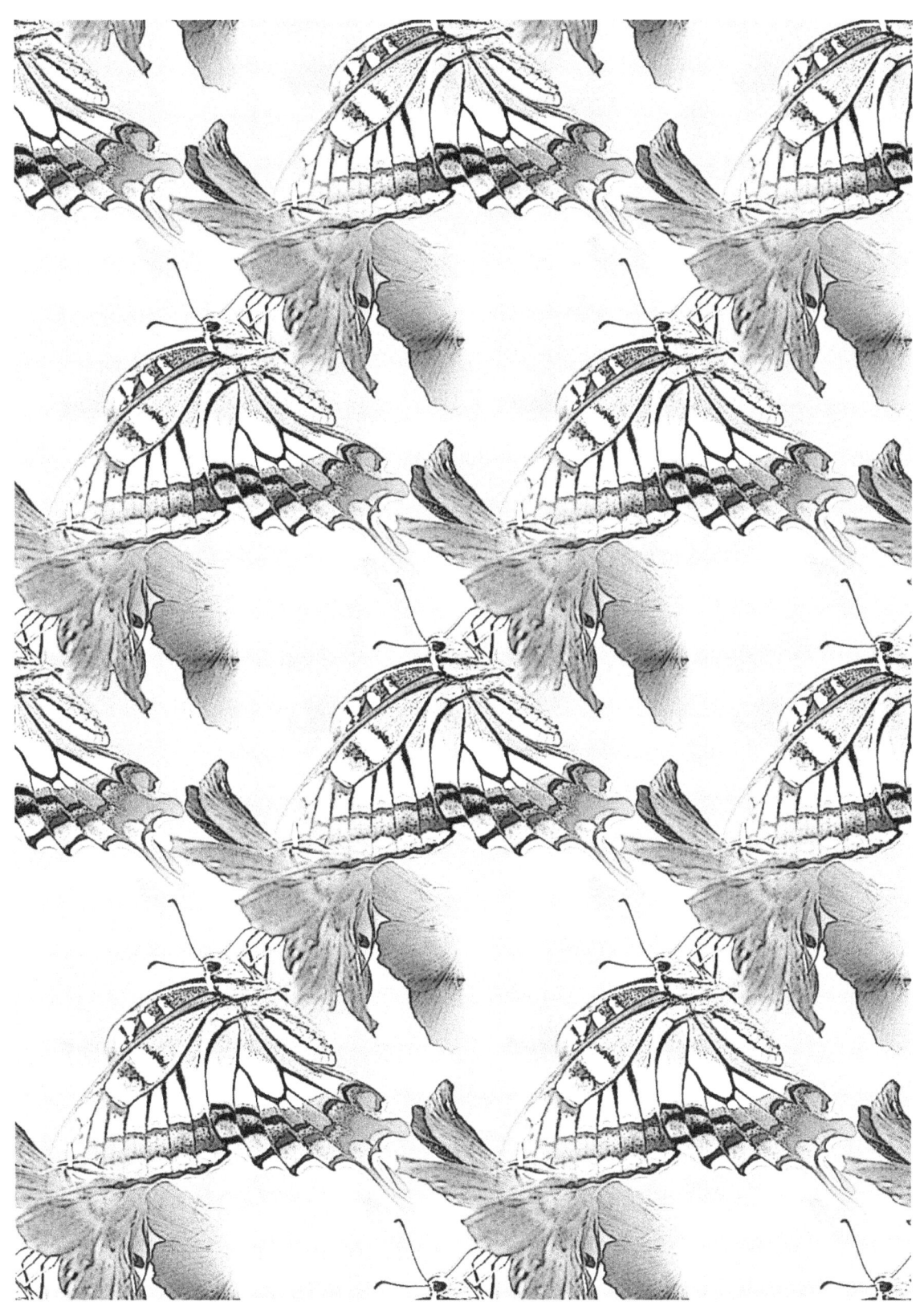

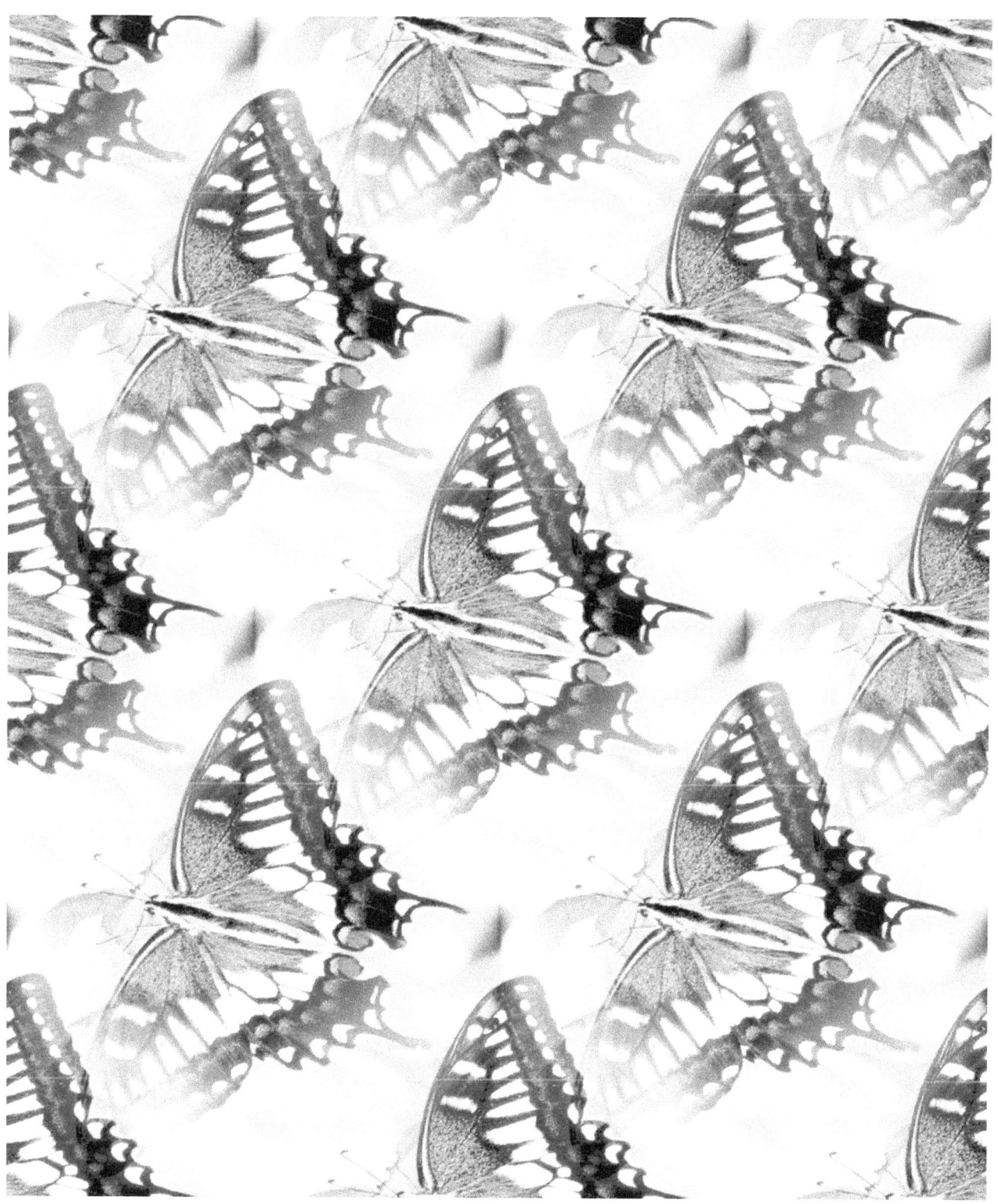

www.ingramcontent.com/pod-product-compliance
Lightning Source LLC
Chambersburg PA
CBHW080610190526
45169CB00007B/2952